A Butterfly's
Favorite Plant

by Mary Clare Goller

Butterflies are insects. A female monarch butterfly lays eggs on a milkweed plant leaf.

Monarch butterflies must have milkweed plants. Let's find out why.

butterfly egg

After about two weeks, a caterpillar comes out of the egg. The caterpillar is in the right place. It's on a milkweed leaf.

The caterpillar only eats milkweed plants.
It nibbles away.

Caterpillars eat milkweed flowers.

The caterpillar eats as many milkweed leaves as it can. It grows a little bigger. But it still needs another milkweed leaf.

Once the caterpillar grows enough,
it attaches itself to another leaf.

The caterpillar's skin hardens. It becomes a chrysalis. The chrysalis [KRIS-uh-lis] hangs from a leaf on a milkweed plant.

After about two weeks, the chrysalis splits open. Out comes a butterfly. Its wings are folded and wet.

The butterfly rests on a milkweed leaf as the sun dries its wings. Then it is ready to fly away.

Some butterflies fly away to a warm place for the winter.

Monarch Butterfly Facts

- A female monarch butterfly lays as many as 500 eggs on milkweed leaves.

- A caterpillar hatches from each egg.

- The caterpillar forms a chrysalis.

- A butterfly breaks out of the chrysalis.

A butterfly dies about three weeks after it lays its eggs.